THE WEAVE OF STONE AND LIGHT

Alchemy, Consciousness and the Sacred Landscape

BY
SIRI HEIMDAL LAGERLØV

GROW BRIGHTER
MEDIA

The Weave of Stone and Light

Alchemy, Consciousness and the Sacred Landscape

1st edition

Created in the United Kingdom

ISBN 978-1-9192155-3-2 (Kindle)

ISBN 978-1-9192155-2-5 (Paperback)

Dedication

To all who have guided and inspired this work. Thank you for showing me how the mundane is woven with the magical.

Disclaimer

This book is a reflection of my personal journey, weaving together my experiences, insights, and what I have found to be my truth. While I've aimed for a grounded and neutral approach, I encourage you to read it with an open mind, remembering that your own intuition is your most important guide.

Please listen to your own needs as you read. Your emotional well-being is the top priority. Take your time and focus on self-care.

The information contained within this book is intended for educational, philosophical, and self-help purposes only. It is not intended as a substitute for professional medical advice, diagnosis, or treatment. Always seek the advice of a qualified healthcare provider with any questions you may have regarding a medical condition. The author and publisher make no claims as to the effectiveness or medical benefit of the techniques described herein. The reader is responsible for how they use the information in this book.

Praise for The Weave of Stone and Light

A Note of Deep Gratitude

I am beyond grateful and deeply honoured to present the following endorsements for The Weave of Stone and Light. The generous support and belief in this project from such respected authors, experts, and masters of modern alchemy and ancient mysteries, many of whom are mentors and friends, is a profound validation of this work. Their words affirm the timeless journey of the Conscious Creator and the powerful resonance of the Sacred Landscape.

"This is a fascinating and timely book that uncovers the hidden design behind ancient British sites like Stonehenge, revealing it as a complex technology of consciousness. Siri's work bravely tackles the historical link between practical alchemy and modern quantum principles, offering a truly radical, and well-researched, explanation for the interconnectedness of matter and mind. Highly recommended."

~Graham Phillips, British Author, Investigative Researcher, and Expert on Ancient Mysteries

"Siri has woven together an original and insightful guide for the modern spiritual journey. By exploring the wisdom held within Britain's sacred sites—from megaliths to ley lines—The Weave of Stone and Light illustrates ways to anchor our personal awakening into the very fabric of the Earth. This book beautifully bridges ancient Hermetic philosophy with the call to be a conscious, grounded participant in Nature's Mysteries, offering us some fascinating techniques and approaches to deepen both our relationship to Nature and Self."

~Philip Carr-Gomm, 'Druid Mysteries', Author, Psychologist, and Former Chosen Chief of The Order of Bards, Ovates & Druids (OBOD)

"Two main interpretations of leys co-exist, one being the 'Old Straight Tracks' of Alfred Watkins of Hereford and his contemporaries. The second is the view that they indicate pathways of energy crossing the land. Siri Lagerløv here provides a coherent retelling of this later view, its ramifications and links to Alchemy and practical ways to apply the knowledge. Highly recommended."

~Adrian Hyde, Secretary, the Society of Ley Hunters

"Siri has produced a timely and significant book that successfully tackles the historical divide between practical alchemy, the mysteries of the Sacred Landscape, and modern quantum principles. She presents the ancient sites—the Stone Temples—not as static monuments, but as complex technologies of consciousness designed to anchor intent. This work is a compelling synthesis, offering a deep, well-researched, and ultimately actionable introductory blueprint for the Modern Alchemist seeking to harness the Earth's energetic matrix for true personal transformation. A valuable contribution to the field."

~Peter Knight, researcher, dowser and author of 14 books about sacred sites and ancient wisdom.

"It's a pleasure to witness Siri bringing forward her wisdom and dedication in the realm of transformation and personal mastery. By weaving the significance of ancient sacred sites and ley lines into her coaching, she highlights a profound truth: there are many different ways to create a life of fulfilment. For anyone seeking to merge Earth wisdom with conscious creation, this book is a valuable and insightful resource. Siri's work offers readers an opportunity to explore new dimensions of personal growth and self-discovery, guided by the mysteries of ancient traditions and the pursuit of purposeful living."

~Hannah Knies, Creative Development Coach & Co-Founder of Magnetic Mind Certification

Table of Contents

About the Author

Siri is a Modern Alchemist, Life Strategist, and Leadership Coach who specialises in guiding spiritual leaders, change makers, and the spiritually curious through profound transformation. After navigating major life curveballs and reinventing herself numerous times, Siri helps individuals reconnect with their core values to design their Life 2.0 with Authentic Confidence, building a stable, unshakeable life they truly love.

Her book, The Weave of Stone and Light, shows how to align inner mastery with the energetic landscape around us. Siri brings a wealth of experience, spanning youth politics, corporate boardrooms, and extensive mastery-level certifications in conscious creation and holistic therapies.

Her deep personal connection to the British Isles and its ancient sites shapes her writing. Her passion for exploring the sacred landscape was sparked by a visit to Stonehenge and the discovery of ley lines. She now lives and works from a property where two ley lines converge, a unique energetic point that inspires her research into the Weave of Stone and Light.

Siri is a speaker who delivers presentations and programmes both online and in person on topics ranging from the journey of conscious creation to achieving Authentic Confidence.

To explore the practical application of the Zero Point, her coaching programmes, or to book Siri as a speaker, please visit her website www.sirilagerlov.com (expected to be live by February 2026).

For professional enquiries, you may contact her directly at contact@coachsiri.co.uk.

CHAPTER 1

THE MODERN ALCHEMIST: WEAVING STONE AND LIGHT

1. The Modern Alchemist: Weaving Stone and Light

The pursuit of alchemy was never a material one. As the great psychologist Carl Jung observed, the true purpose of the alchemists was a profound metaphor: the transformation of the self from a state of raw, chaotic "lead" into the brilliance and clarity of "gold." This is the ultimate quest for inner mastery, and it is just as vital today as it was centuries ago.

The modern alchemist, therefore, is engaged in a far grander work: the transformative alignment of their inner and outer worlds. This is the Weave of Stone and Light.

The Stone represents the matter of our existence—the physical body, the grounded reality of the Earth, the history we carry, and the tangible world we observe. This is the lead that must be purified.

The Light represents consciousness, the cosmic energy, the quantum potential, and the boundless, non-linear flow of the universe. This is the gold—the refined awareness that allows us to collapse potential into physical form.

True mastery lies in learning how to weave these two elements together, guiding the Light of your intention to reshape the Stone of your reality.

Inner Alchemy: The Cleansing Practice

The process of moving from lead to gold isn't about magical spells; it's a dedicated practice of self-awareness and intentional action. Long before Hermeticism, mystics of the

Sufi tradition embarked on a similar quest for inner purification (tazkiyat al-nafs), knowing that true reality is found through a deep connection to the unseen world.

The difference between chaos and clarity is a conscious choice—adjusting the lens, focusing on the light, and bringing your full picture into vibrant clarity. It is the difference between a stagnant pond and a flowing river.

Practice: Purifying the Vessel of Consciousness

Before you begin this journey, let's try a simple visualisation to cleanse your inner state and prepare for the transformation ahead. This practice, familiar in all true schools of spiritual healing, is based on the alchemical process of purification.

Find a quiet place to sit comfortably. Close your eyes and take a few deep, intentional breaths. As you settle into your body, imagine yourself as a clear glass bottle. Inside this bottle, there is water. It might be murky, cloudy, or filled with the debris of a busy day—representing the state of "lead" you may be carrying.

Now, above your head, visualise a point of brilliant, white light. This light is pure, expansive, and filled with energy from the universe. As you exhale, imagine gently releasing some of the murky water from the bottom of your bottle, letting it drain away. As you inhale, draw in the pure, bright light from above your head, allowing it to fill the empty space inside.

Continue to breathe in the light and exhale the murkiness. With each breath, the water in your bottle becomes clearer, brighter, and more vibrant, until the murky water is completely replaced by pure, radiant light. Feel the clarity and peace this brings to your mind and body. You have now performed an act of inner alchemy.

This simple practice of purifying your inner world can be a subtle key to unlocking your potential as a conscious creator.

When Ancient Wisdom Meets Quantum Science

For centuries, alchemy was dismissed as a relic of a pre-scientific age. Today, the profound parallels between these ancient ideas and the latest discoveries in physics are impossible to ignore. Many modern physicists find that the Hermetic principles that formed the core of alchemical thought sound remarkably like the theories they are exploring today.

The "magic" of the alchemists is finding an elegant reflection in the intricate, often bizarre, world of quantum mechanics:

The Observer Effect: The alchemist was never a passive spectator; they were an active participant whose consciousness was a critical ingredient for transformation. Quantum physics offers a parallel: the act of a human observer forces a particle to collapse from multiple possibilities into a single reality. We will explore the scientific proof that your Light is a catalyst for the Stone in Chapter 4.

Vibrational Resonance: By consciously raising our personal vibration through practices like meditation and mindfulness, embodiment and rewiring of neuro pathways, we can align ourselves with the frequencies of what we desire. This isn't magic; it's a principle of vibrational resonance. Like a tuning fork, you attract what you are in harmony with. This is how we become conscious creators who attract what we desire like magnets—not by willing something into existence, but by becoming a vibrational match for it.

The journey from lead to gold, from chaos to clarity, is a journey of understanding and mastering our own energetic frequency. When we learn to do this, we are truly honouring the ancient alchemists and their most profound discoveries.

The Mirror of the Universe: Microcosm and Macrocosm

Central to alchemical and Hermetic thought is the profound principle of the Microcosm and Macrocosm. This is the cosmic truth captured in the ancient axiom, "As above, so below; as below, so above."

The alchemists viewed the human body and spirit as a miniature cosmos, containing the same fundamental forces that govern the vast expanse of the universe. Understanding this mirroring allowed the alchemist to work on themselves as the key to understanding the greater cosmic order.

This concept resonates deeply with the sacred sites we will explore. An ancient stone circle, aligned with the rising sun on the equinox, is a perfect physical reflection of the cosmos—a place where the Macrocosm can be experienced on a human scale. This book is an invitation to explore your own inner world as the key to the universe itself.

The Ultimate Treasure: The Philosopher's Stone

Lastly, we arrive at the enigmatic Philosopher's Stone. For the true alchemist, this ultimate object was a metaphor for the ultimate treasure within ourselves, a state of enlightenment, wholeness, and perfection. The search for the Stone is the journey of integrating all aspects of yourself: harmonising with your shadows, resolving inner conflicts, and transforming the raw, unrefined parts of your inner world into a cohesive and brilliant whole.

The Stone is the profound resonance you can achieve when your microcosmic self is fully aligned with the greater macrocosm.

My Personal Alchemical Path

My own journey into this form of alchemy began long ago, far before I understood the term. As a child, I was fascinated by the books of Erich von Däniken, spending hours looking at photos of ancient cultures and imagining a world of aliens and spaceships. In my early teens, my mother introduced me to the work of Jannicke Jarlum, a Norwegian spiritual author and artist, and I found myself reading her book over and over, hungry for more. As I approached my late teens, when in spiritual settings like meditation circles, I would often draw attention to the intensity of my heart chakra, an early sign of the energies I would later learn to work with and to which capacity I was quite oblivious to at the time.

Up to my mid-twenties, my spiritual path was nearly a hidden, nearly denied passion. My dream of becoming an author was on the shelf, I had lost the structured direction in my life.

Nevertheless, the next part of my life took me through tuition in Shamanism, Druidry, Sound Healing, Reiki Mastery, Crystal Healing and Spiritual Healing to mention some. It wasn't until my thirties, over a cup of coffee with a friend, that I was first introduced to Hermeticism and the book The Kybalion. My friend hinted that I wasn't in the right place to read it, which, of course, was all the challenge I needed. I bought the book immediately. While not an easy introductory volume, it ignited a lifelong passion and was a foundational step in my own alchemical journey.

My forties took me into transformational coaching and the marvels of quantum science.

Contrary to concepts like Nirvana and Satori, often described as sudden revelations of "striking the gold," I find the journey of inner alignment is a sustained, deliberate process; yet, the moment of conscious understanding certainly feels sudden. Like Mikao Usui, who developed Reiki after a 21-day mountain retreat, the dramatic revelation is the culmination of years of preparatory work. For me, the journey has been gradual and taken sudden and unexpected turns but has exponentially become more streamlined and aligned—until I looked back and could recognise that "Hey, this is what it looks and feels like." That moment I would think most certainly feels similar to touching the top of Mount Everest. However, it is a journey and to stay in alignment can be hard when resistance shows up, although it gets easier when you know what resistance really is. The reason why this journey of inner alignment often takes such a sustained time to culminate in "gold" is that through the process of individuation (age 3-7), also called developmental imprint process, we get so conditioned into the "lead" state that we need to reteach our brain to change. I will speak some more about this process later, and it is also discussed substantially in another book that is in the pipeline.

This exploration was a confirmation that what I had learned from my early teachers was not just theory, but a living, breathing reality I could feel and experience. It was a clear sign that the sacred and the material are inextricably linked.

I believe we can all learn to master this process. This book is your guide for start of your journey. We'll start by diving into the ancient principles of alchemy and Hermeticism, and then we will walk through the sacred stone circles of Avebury and Stonehenge, which I consider to be ancient temples of alignment. By the end, I hope to show you that the "magic" of these places and the power of alchemy are not things of the past but are alive in you and in the world right now.

To begin this journey into the tangible reality of our connection to the Earth, we now turn our attention to the physical landscape. By exploring ley lines and their powerful, unseen energies, we will begin to see that the Earth itself is a living organism, a cosmic body with its own life-giving circulation.

CHAPTER 2

THE WEB OF LEY LINES: THE EARTH'S ENERGETIC ARTERIES AND VEINS

2. The Web of Ley Lines: The Earth's Energetic Arteries and Veins

Throughout our exploration of alchemy and consciousness, we'll keep coming back to the profound idea that our inner world is a microcosm of the vast macrocosm of the universe. When we look at the Earth itself, we see the Weave of Stone and Light in action. The land is not just inert matter (the Stone); it is a living, breathing entity with its own energy (the Light), a network of powerful channels often referred to as ley lines.

What Are Ley Lines?

In essence, ley lines are invisible, straight alignments in the landscape, often marked by ancient sites, natural features, or historical monuments. While the term was coined in 1921 by Alfred Watkins, an amateur archaeologist and photographer, his original observation was that these sites lined up in a straight path, as if they were ancient trackways. The term "ley" comes from Anglo-Saxon, meaning a clearing or open grassy land. Modern geomancers and sensitives have since discovered that these alignments often correspond to powerful, flowing currents of energy, the very arteries and veins of the Earth.

Think of ley lines as the planet's own circulatory system, a vast and intricate web of light that carries subtle, life-giving properties. They are not merely physical lines on a map, but a profound energetic reality that connects sacred sites, hilltops, and springs across the globe.

Ley Lines are not to be confused with Earth energy lines. Ley lines are the concept of straight, deliberate alignments of ancient sites, while earth energy lines are considered the naturally occurring, often the Caduceus-like invisible flows of subtle energy that are believed to exist in the landscape.

Where Do They Come From?

The question of ley lines' origin leads us into the heart of telluric (of the Earth, Tellus) energy. Just as our bodies have an energetic field, so too does the Earth. As we've already seen with the Earth Chakras, and as we can find in meridians and energetic lines on our own bodies, the planet has its own circulatory system. Ley lines are believed to be channels for this energy, known as telluric flows. These telluric flows are energetic streams, paths, or trajectories that exist near the surface of the Earth. Scientists recognise that telluric currents, which are influenced by the planet's magnetic field and solar activity, arise from natural electrical resistance and induction within the planet's crust.

Ley lines, however, appear to carry a more subtle energy, operating with properties that differ from electricity and magnetism. We can think of them as the Earth's meridians. From this perspective, the sacred sites that appear along these energetic pathways act as powerful energy hubs, similar to acupressure points on the human body. These are concentrated vortices where energy can be accessed, released, or amplified for healing and alignment. While the term 'currents' is often used to describe any kind of flow, for the purpose of understanding ley lines, we are speaking of energetic pathways that cannot be measured with current scientific instruments. Yet, this debate is at the very heart of the journey from "lead" to "gold." It is the choice to quiet the rational mind and trust our intuitive self to experience the unseen.

This earthly energy is a part of a much larger cosmic system, too. In the same way that ley lines form a planetary energetic web, cosmologists speak of the Cosmic Web—a large-scale structure of dark matter filaments and galaxy clusters that forms the very backbone of the universe. This shows us that the pattern of energetic interconnectedness that we find on Earth is a universal principle, echoing the Hermetic axiom of "as above, so below."

While Watkins' original theory was based on visual alignments, a more holistic understanding suggests that our ancestors, through their intuitive knowledge and sensitivity to the land, built their most sacred structures—the stone circles, the mounds, and the ancient temples—on these powerful energetic points. They knew how to feel and perceive the flow of energy in the landscape, and they created these markers to honour and perhaps even harness that power.

How Are They Measured?

For centuries, people have used a practice called dowsing to detect these and other energies. Using simple tools such as L-shaped metal rods, a pendulum, or a forked stick, dowsers walk the land and pay attention to how the tool responds.

It is important to note that there is no instrument to measure the energy of ley lines in conventional scientific units. The only "instrument" that can detect them is the human body itself. Dowsing is not a magical trick; it's an energetic dance. It requires the dowser to quiet their mind and enter a state of inner alignment—that zero point we talked about. The dowser is consciously using their Light (consciousness) to perceive the currents flowing through the Stone (Earth). The dowsing rods act as an extension of your body, amplifying your innate sensitivity to these energies, and training you to access your intuitive receiver on demand.

Why Are People Hunting Them?

People hunt ley lines for several reasons, but at the heart of it, the quest is for healing, alignment, and connection. Ultimately, ley line hunting is an act of modern alchemy—the process of turning unseen, invisible energy into a tangible experience of healing, connection, and profound transformation.

My personal theory is that sensitive people back in the day would use these lines for navigation from one location to another simply by following the energy lines. They understood that by marking these powerful, straight corridors of energy with monumental sites—from megalithic circles to ancient avenues—they were creating an intentional, robust roadmap across the land, connecting sacred centers of power. This ancient understanding extends beyond just stone; I believe that sensitive people observed nature following these lines too. It is no coincidence that you often find rows of mighty oaks following these alignments, suggesting the trees were planted there as landmarks for travellers along these energetic routes.

My first experience with dowsing rods

My first experience with dowsing rods was at a local, hidden temple. It was a brand-new set of rods bought in Avebury, and I had done a meditation with them first to connect and become sensitive to holding them. It took a little while to get the feel of holding them correctly, which is loosely, to allow them to move freely. I think everyone tries to manipulate their movement at first, but there's no need for that.

I still remember the first time I felt the rods moving on their own in my hands. I was walking on the path towards the temple, and they strongly pointed forwards, changing direction towards the earthworks as the path twisted and turned. I had been shown the map where all the local ley lines were scribed in, and I instantly recognised we had to be on a strong line.

As I entered the grove that conceals the temple's earthworks, I came to the point where the entrance path met the old path that encircled the pond and mound. The rods started gently turning like mini rotor blades, much as if I were on the Magnetic North Pole. I had brought a compass, and even the needle on the compass was spinning around. I stopped, closed my eyes, and took a deep breath, and it felt like I was standing in an energetic portal. It was an amazing experience that I will always remember. In that moment, I felt a deep sense of awe, like I was a witness to something truly sacred and holy—and a small part of it myself.

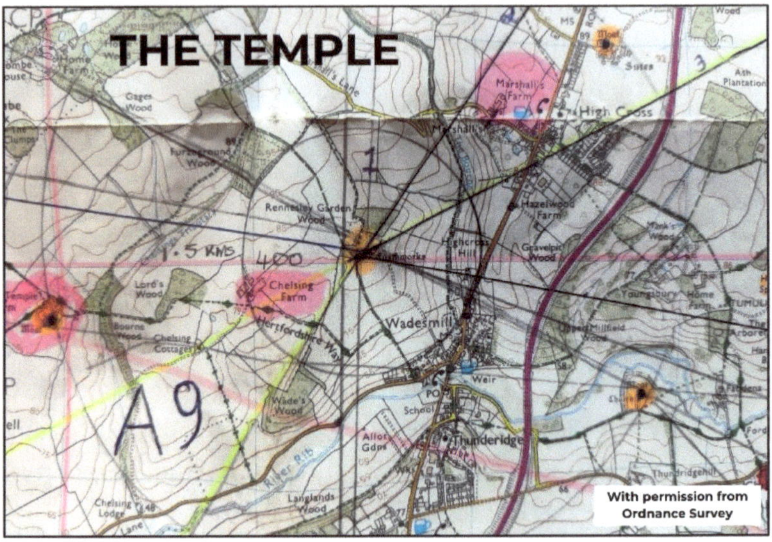

Earth Chakras: The Planetary Energy Points

As we talk about ley lines and their role as Earth's arteries and veins, we cannot overlook the concept of Earth Chakras... These points are thought to form at the intersections of planetary energy lines—the very ley lines we are discussing. The most relevant example for our journey is the Heart Chakra (Anahata), believed to be located at Glastonbury and Shaftesbury, England. This connection to the spiritual heart of the planet reinforces the idea that the entire energetic grid facilitates healing, love, and compassion.

Below is an illustration of the Earth Chakras on a world map as an overview.

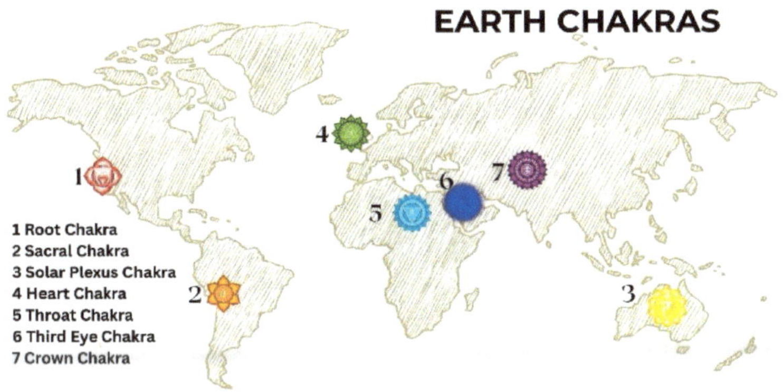

[Picture description: Root Chakra (Muladhara): Mount Shasta, California, USA. This is considered the core of Mother Earth's energy. Sacral Chakra (Svadhisthana): Lake Titicaca, on the border of Peru and Bolivia. This location is associated with creativity and new life. Solar Plexus Chakra (Manipura): Uluru and Kata Tjuta, Australia. These are vital points for global health and vitality. Heart Chakra (Anahata): Glastonbury and Shaftesbury, England. This site is known for its energy of love, healing, and compassion. Throat Chakra (Vishuddha): The Great Pyramid of Giza and Mount Sinai in Egypt. This is sometimes called the "Voice of the Earth". Third Eye Chakra (Ajna): This chakra is considered a "floating" location, sometimes associated with the Mount of Olives in Israel or Glastonbury, England, as it may shift with new spiritual ages. Crown Chakra (Sahasrara): Mount Kailash in Tibet. This mountain is known as the "Crown of the Earth" and is a place of sacredness and spiritual journeys.]

Ley Lines in Action: Avebury and Stonehenge

While the concept of ley lines may seem abstract, the sacred sites of Avebury and Stonehenge stand out as prime examples, acting as a testament to the ancient understanding of Earth's energies.

At Avebury, the West Kennet Avenue itself is considered a ritual path energetically connected with the famous St. Michael and St. Mary lines... Just a short distance away, Stonehenge is believed by many to be a focal point for an even greater concentration of power. This convergence is believed to be why the site was a place of such profound importance to ancient peoples, possibly for healing or spiritual ceremony.

Yet, this debate is at the very heart of the journey from "lead" to "gold." It is the choice to quiet the rational mind and trust our intuitive self to experience the unseen. We may not have a scientific instrument to measure the energy of these places, but millions of people, from ancient times to the present day, have felt it.

Chapter 2 Summary

In this chapter, we have journeyed beyond the familiar to explore the concept of ley lines as the Earth's energetic arteries and veins. We've seen how these invisible pathways... are a tangible reality that can be felt through the intuitive practice of dowsing. Ultimately, ley line hunting is the modern alchemist's way of weaving the Stone (Earth) with the Light (Consciousness), translating unseen energy into an experience of healing, purpose, and profound connection to the living world around us.

CHAPTER 3

A MODERN TAKE ON ALCHEMY: INNER MASTERY

3. A Modern Take on Alchemy: Inner Mastery

Casanova – the Alchemist of Attraction

Building upon our understanding of the Philosopher's Stone as a metaphor for inner mastery and alignment, let's now turn to a truly unexpected figure: the legendary lover, Giacomo Casanova.

Born in Venice in 1725, Casanova was an Italian adventurer known for using pseudonyms and claiming alchemical knowledge to gain favour. For him, alchemy was a tool for social climbing rather than genuine spiritual practice. Yet, if we choose to see him in a different light, his life offers a powerful, if unconventional, example of how internal state influences external reality.

Casanova's extraordinary success in matters of the heart likely appeared as some inexplicable charm, a "weird psychic trick" to irresistibly attract whomever he desired. This is the core lesson: I often say to those I coach that becoming a true master of your superconscious field isn't about forced positivity or sheer willpower; it's about inner mastery. Casanova's consistent success was the outward manifestation of a profound internal process.

This is a mastery of the alchemical 'Great Work' known as solve et coagula—to dissolve and to coagulate. Casanova was able to 'dissolve' his insecurities and self-doubt, not through willpower but through a deep, embodied belief in his own magnetic presence. He then 'coagulated' this refined energy into a powerful external reality that attracted exactly what he desired. His charm was not a façade; it was the outward manifestation of a profound internal process that anyone can learn.

The Inner Work: Purification and Alignment

The first part of our journey is to understand how our personal reality is formed. The process of individuation, a term popularised by Carl Jung, is the unconscious programming that happens in our formative years when we create "rules" for ourselves based on our need to belong. These patterns, often formed as early as the toddler stage, can rule us for the rest of our lives without our conscious awareness.

This is where true alchemy begins. The alchemical process is the conscious journey of confronting and engaging with these deep-seated patterns. A key part of this is our shadow, which refers to the parts of our personality we unconsciously repress. These aren't just our inner critics; they also hold our dormant power, untapped talents, and brilliance. In many cultures, the internalised 'who do you think you are?' ethos acts as a massive brake, teaching us to repress our Light and not stand out. Reclaiming this repressed potential is the essential work that transforms our whole being into gold.

True alchemy, the way I see it, is the process of healing and learning to handle these patterns and to integrate the shadow. This is the purification of the lead—the essential, difficult work that transforms our whole being into gold. When we reclaim our shadow, we are in alignment with our soul's desire, and we silence the inner voices that talk us out of living our true nature and purpose. This is the critical first step to becoming a conscious creator.

A Conscious Creator is someone who applies the laws of Personal Alchemy to their life, for instance using the Hermetic principles to achieve mastery. This mastery involves three key recognitions:

Quantum Responsibility: They recognise the quantum principle that consciousness collapses potential into reality. They accept responsibility for the energetic output they are broadcasting (thoughts, beliefs, and emotional state).

Internal Blueprint: They actively work on transforming the internal "base metal" (limiting beliefs, fears, and unconscious programming) into "gold" (purpose, destiny, and success).

Grounded Intent: They understand the power of anchoring their intentions into the physical world, using methods—such as engaging with the stable energetic grid of the Sacred Landscape—to stabilise their Superconscious connection and manifest their desired reality.

In essence, the Conscious Creator is the modern Alchemist, transforming their inner world to transform their outer world, using the Earth's energy as their laboratory.

The Three Levels of Consciousness: A Map for Your Inner World

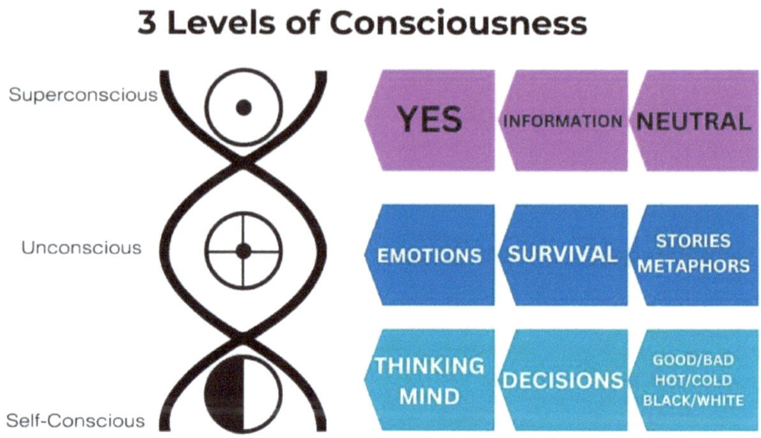

To better understand your personal alchemy, it is helpful to look at the three levels of consciousness: the Self-Conscious, Unconscious, and Superconscious. Think of them as the three main actors in your inner world, each with a very specific role to play. You can imagine this in a metaphor of a ship:

The Self-Conscious Mind is the Captain: It does your conscious thinking, makes decisions, and is busy making sense of your external world based on polarities (right/wrong, good/bad). It is the logical processor.

The Unconscious Mind is the Crew: It is the keeper of your emotions, survival instincts, and all your past experiences. It runs all the operations that your conscious mind isn't even aware of, creating patterns and coping mechanisms to keep you safe. These rules are the result of your Developmental Imprint Process—the unconscious programming that often dictates your reality until you choose to rewrite it.

The Superconscious Mind is the Guidance System: It is the wellspring of intuition, creativity, and your innate spiritual knowing. It holds a bigger, more expansive perspective and is always neutral and unbiased.

The process of inner alchemy is about deliberately aligning the Captain (Self-Conscious) with the Superconscious Mind (the ultimate guidance system) so that the Crew can finally execute a cohesive, powerful vision for your life.

For this to happen, the unconscious needs a clear destination. Think of it like this: to get to any destination, whether by sea, land, or mind, you need a starting point ("A") and a clear goal ("B"). This destination is determined by your Self-Conscious Mind (your wishes, desires, and goals for the particular journey in your life). The Unconscious Mind acts like your internal GPS, ready to plot the most efficient route. However, if your programming is stuck in "ice"—those fixed limiting beliefs—it acts as a faulty GPS full of "bugs," constantly rerouting you back to an old, familiar location, or simply refusing to start the journey.

This is why the process of inner alchemy is the living application of solve et coagula: Think of "ice" as a fixed belief or pattern. Through dissolving (solve), you release this fixed belief, allowing it to become a fluid, unlimited state of potential. Then, through coagulating (coagula), you reform this energy into a new, more aligned "fixed" state that serves your highest good, giving the Captain and Crew a powerful new, reliable map for life.

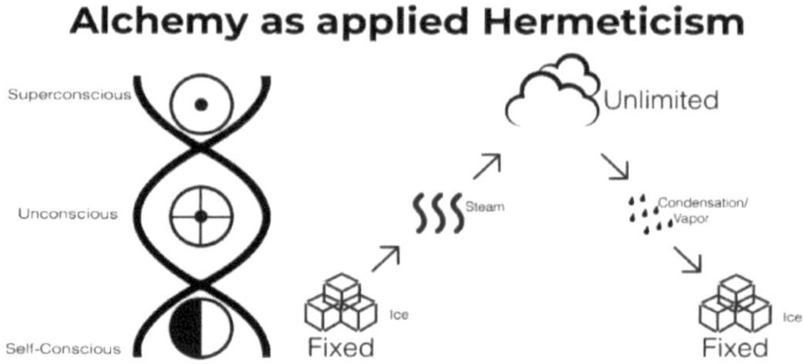

Chapter 3 Summary

In this chapter, we explored the idea of inner alchemy through the lens of Casanova's internal mastery. We saw that his secret lay in the active process of solve et coagula—dissolving old patterns and coagulating new ones. This led us to our own work, beginning with the purification of the lead through individuation and shadow work. We then mapped out the tools for this work by learning about the distinct roles of the Self-Conscious, Unconscious, and Superconscious minds.

We have seen how inner alchemy is the process of transforming our personal reality by mastering our own mind. But our mind is not an isolated kingdom; it is a part of the vast, interconnected fabric of the universe. In the next chapter, we will expand our perspective beyond the personal to explore the universal principles that govern both the outer world of matter and the inner world of consciousness. We will connect the transformative power of alchemy to the ancient wisdom of Hermeticism and its Seven Principles and the modern insights of quantum physics, revealing how these seemingly disparate fields are, in fact, two sides of the same coin.

CHAPTER 4

A QUANTUM PERSPECTIVE THROUGH THE LENS OF HERMETIC PRINCIPLES

4. A Quantum Perspective Through the Lens of Hermetic Principles

Building on our modern understanding of alchemy, we now turn to a perspective that bridges the ancient and the new: the Hermetic principles. These seven core tenets, such as "As Above, So Below" and "The Principle of Vibration," formed the philosophical backbone of alchemical thought. They describe a universe that is fundamentally interconnected, where everything is in constant motion and governed by unseen energies.

Well then, who were the Hermetics? The Hermetics were followers of a philosophical and religious tradition known as Hermeticism, based on writings attributed to Hermes Trismegistus ("Thrice-Greatest Hermes"), a legendary Hellenistic figure. This figure is a syncretic combination of the Greek god Hermes and the Egyptian god Thoth, both deities associated with wisdom, writing, and communication.

Originating in Hellenistic Egypt between the 1st and 3rd centuries CE, this was a vibrant cultural melting pot where Greek, Egyptian, and other Near Eastern ideas intermingled. The Hermetic writings, known as the Corpus Hermeticum and the Emerald Tablet, were attributed to this mythical sage. While not a historical person in the traditional sense, Hermes Trismegistus became a symbol of ancient wisdom. The core beliefs explored in these texts include the nature of God (often referred to as "The All"), the cosmos, and the relationship between the spiritual and material realms.

Hermeticism also introduced the idea of *Prisca Theologia (Primal Theology)*—that a single, true theology existed in antiquity and that various ancient wisdom traditions shared fundamental truths. Hermetic principles had a significant influence on Western esoteric traditions, including alchemy, astrology, and later, the magical practices of the Renaissance.

PRISCA THEOLOGIA

This is an ancient Hermetic idea that there exists a single, true theology or wisdom that was given to humanity in antiquity. This "Primal Theology" is said to be the source from which all other major wisdom traditions, from Hermeticism to Platonic philosophy, arose. It suggests that while various traditions may use different languages or symbols, they all share a fundamental, universal truth at their core. In Hermeticism, Hermes Trismegistus is considered to be the ultimate prophet of this theology, making the concept a cornerstone for anyone seeking to unite different spiritual paths.

Modern science, particularly through the framework of quantum physics, has begun to provide a language for these same concepts. While the ancients spoke of a living, vibrating cosmos, quantum mechanics reveals a subatomic world where particles exist in multiple states at once and are influenced by the very act of observation. This strange and beautiful reality resonates deeply with Hermetic wisdom, suggesting a universe far more fluid and malleable than traditional classical physics would allow. The ancient alchemist's belief in the power of conscious intent and inner state to influence the outer world is now finding a surprising parallel in the quantum realm.

Seven Hermetic Principles

There are seven Hermetic Principles that form the basis of Hermetic philosophy. These principles are not laws in a scientific sense, but rather universal truths that describe how reality works from a spiritual and energetic perspective.

Understanding them gives us a powerful framework for seeing the world differently.

It is important to understand that these principles do not exist in a hierarchy; no single law is more important than another. They are foundational, universal truths, much like the seven notes of a musical scale—no single note is more important, but their unique combination creates the music. The seven Hermetic Principles work in concert with one another. To truly master one principle, you must understand how it relates to and interacts with all the others. This interconnectedness is the source of their profound power.

The Art of Transmutation

The true power of the Hermetic Principles lies in their use as tools for transformation. Alchemy, in its deepest sense, is the art of transmuting one state into another by intentionally applying these principles. For example, a person cannot eliminate the Principle of Polarity, which dictates that every emotion has an opposite. However, they can use the Principle of Mentalism to work with it. By consciously choosing to hold a thought or a feeling on the opposite end of the spectrum (say, peace instead of anxiety), one is performing a form of mental alchemy, shifting their experience along the same energetic line. This is the ultimate goal of Hermeticism: not to defy the laws of the universe, but to understand them so completely that you can use them to consciously shape your own reality.

The Planes of Principle

While all of the Hermetic Principles are equally important, their work often takes place on different planes of reality. In the ancient texts, these are sometimes referred to as "higher" and "lower" principles, but this is not about a hierarchy of significance. Instead, it describes a relationship of cause and effect.

The higher principles, such as Mentalism, are considered the most fundamental and abstract. They are the initial causes or blueprints in the mental realm. The lower principles, such as Rhythm or Polarity, are those that manifest in the physical, emotional, or material realms.

To explain further: your thought (a higher principle) is the original cause, and the emotional feeling or physical action that results is the effect, manifesting on a lower plane. By understanding this relationship, you truly grasp how to use one principle to influence another and become a conscious creator.

The Principle of Mentalism: "The All is Mind; The Universe is Mental."

This principle states that all of existence—everything we see, feel, and experience—is a mental creation within a vast, universal consciousness known as "The All." It suggests that the physical world is a reflection of thought and that our own minds are a part of this greater cosmic mind. The universe is not a collection of random parts, but a single, living thought.

The following is a practical modern-day example of the Mentalism Principle: imagine you're training for a marathon run. Your physical preparation—the long runs, the strength training, the nutrition—is certainly vital, but the mind plays an important part for your performance in the race, whether you fulfil your goal or not. The conscious mental blueprint of you crossing the finish line, the inner belief that you can and will do it, becomes the cause that powerfully shapes your physical effect. It's the mental visualisation that precedes and guides the physical manifestation. This is the essence of Mentalism: your internal world of thought and intention directly informs and creates your external reality.

The very creation of both Avebury and Stonehenge began with specific mental blueprints held by their builders. The scale and complexity of Avebury suggest an expansive conceptualisation of a sacred landscape, while the precise geometry and alignments of Stonehenge point to a focused and deliberate intellectual framework. Both sites stand as testaments to the power of intention and the ability of ancient minds to manifest complex visions in stone and earth.

The Principle of Correspondence: "As above, so below; as below, so above."

This principle is one of the more common of the Hermetic principles and means that there is a constant mirroring between the different planes of existence, whether they be physical, mental, or spiritual. What happens on a large scale (the macrocosm, like the cosmos) is reflected on a smaller scale (the microcosm, like humanity and the Earth). This allows us to understand the bigger picture by studying the smaller one, and vice versa.

The second principle invites us to see mirroring relationships between celestial and terrestrial realms. The potential mirroring of constellations or planetary movements in the layout of Avebury, and the clear solar alignments at Stonehenge, both suggest a deliberate attempt to align these earthly structures with cosmic order. Both sites may have been intended to facilitate a connection between the human microcosm and the macrocosmic energies of the universe, albeit through different expressions. When we stand within the stone circles, are we not experiencing the “As Above, So Below” in a tangible way, feeling the echoes of celestial energies beneath our feet?

On a more practical level, think of it as a game of observation within your own life. When you coach a team leader and their team seems overwhelmed and chaotic, what do you see? Often, the leader's internal state—their own stress and uncertainty (the microcosm)—is being mirrored in the larger collective energy of the team (the macrocosm). The chaos you observe on the surface is a direct reflection of the leader's inner world. By helping the leader find calm and clarity within themselves, you can observe a corresponding shift and stabilisation in the entire team. This is "As Above, So Below" at work: the inner world of one person directly influences the outer world of their collective.

The Principle of Vibration: "Nothing rests; everything moves; everything vibrates."

This principle posits that everything in the universe is in a constant state of motion and has its own unique vibrational frequency. From the densest matter to the most subtle thought, everything vibrates. The differences we perceive between things are simply a matter of their different rates of vibration.

The placement of these massive stones, on significant ley lines, suggests an understanding of energetic flow, and assisted by either sensing it, or feeling it, or dowsing with willow sticks. Perhaps the far-reaching circles of Avebury were designed to interact with broader energetic currents across the landscape, while the focused arrangement of Stonehenge amplified specific telluric or celestial vibrations at its central point. Both sites, in their unique ways, could have been intended to modulate and harness the inherent energetic nature of their locations.

A profound, modern-day example of this principle in action comes from the work of Japanese scientist Masaru Emoto. In his experiments, he exposed water to different words, thoughts, and music, and then froze it to examine its crystalline structure under a microscope. He found that water exposed to positive, loving words like "thank you" or "love" formed beautiful, intricate crystal patterns. Conversely, water exposed to negative words like "hate" or "evil" formed chaotic, distorted, or ugly formations. His work, while debated in scientific circles, serves as a powerful metaphor for how a vibration—whether it comes from a sound wave, a thought, or an emotion—can influence and shape the physical world around us. It begs the question: if our thoughts can change the structure of water, which makes up more than 70% of our bodies, what is the transformative power of our inner world on our own being?

THE ENERGETIC INFLUENCE OF WORDS

Thus, on a personal level, this principle applies to us as well. Each of us carries a unique energetic vibration that shifts with our moods and emotional states. The high-frequency vibration of joy and gratitude, for example, feels vastly different to the low-frequency vibration of anxiety or sadness. As a conscious creator, you can learn to wilfully change your vibration. This isn't about ignoring difficult feelings, but about choosing to cultivate a new state, for instance happiness. Furthermore, we can do the deeper work of healing the trauma wounds and triggers that make us so susceptible to those lower-vibrational states, allowing us to stay in a state of inner alignment more consistently.

The Principle of Polarity: "Everything is dual; everything has poles..."

This principle states that everything in the universe has an opposite. Light and dark, hot and cold, love and hate—these are not two separate things, but rather two ends of the same spectrum. Everything exists in a state of duality, and by understanding this, we can move from one extreme to the other. Our conscious self, also called ego or Self-Conscious, operates from this principle.

To understand this on a personal level, consider a time when you experienced a significant setback—perhaps a project at work that failed, a creative endeavour that didn't take off, or a personal goal you didn't reach. In that moment, it's easy to feel defeated and to label the experience a complete "failure."

The Principle of Polarity invites us to pause and realise that "failure" is just the opposite pole of "success." They are not separate realities but two ends of the same spectrum. By consciously choosing to shift our perspective, we can move along that spectrum. What if we reframe the disappointment not as a definitive end, but as a crucial "learning opportunity"? When we do this, the negative emotions begin to transmute. We can see the lessons learned, the resilience we built, and the new direction that has opened up precisely because of the initial setback. This is the profound, personal alchemy of turning leaden disappointment into golden wisdom. A coach often helps a client through this very process, guiding them to see the inherent growth within every challenge and to mentally transmute a negative emotional state into a more positive one.

Consider how the contrasting structures of Avebury and Stonehenge might represent a balancing of energies. The open, encompassing nature of Avebury could be seen as a 'feminine' or receptive pole, while the more defined, focused structure of Stonehenge might represent a 'masculine' or active pole. Perhaps the ongoing academic debate about their true purpose is a perfect example of the Principle of Polarity in action: two seemingly opposing interpretations that might hold different facets of a larger, cosmic truth.

The Principle of Rhythm: "Everything flows, out and in; everything has its tides..."

This principle explains that everything has a natural rhythm and cycle. It's the law of cycles, ebb and flow, rise and fall. Like the tides of the ocean, everything in the universe follows a predictable pattern of rhythm and return. To master this principle is to work with these cycles rather than against them.

The cyclical nature of time and celestial movements is evident in the solar alignments of Stonehenge and likely embedded within the broader design of Avebury. Both sites, in their own ways, may have been used to mark and celebrate these cosmic rhythms, facilitating a human connection to the ebb and flow of energy throughout the year. Avebury's less defined structure might reflect a more continuous engagement with these cycles, while Stonehenge's precise alignments focused on key turning points.

These cycles can also be described by the seasons of the year. For most of the planet it is true that we have seasons with a Winter where nature hibernates, roots grow strong and nurtures itself getting ready for a Spring where plants germinate and start growing, animals procreate and nature changes colours. During a Summer new life emerges from more seeds germinating and animals welcome their offspring, an abundance of foods for gathering. During Autumn, annual plants die off, perennial plants and a several species of animals get ready for hibernation, game is ready for hunting and traditionally the farmers harvest and create abundance for us humans and also help prepare us for the winter months where fewer things grow. For instance, if you're a creative person, in your work or privately, look through to the last years through a retrospective lens and notice what time of the year are you most creative? What time of the year do you execute your ideas? What time of the year do you hibernate and let your ideas grow? When do you tend to nurture yourself more? Doing an analysis of this for yourself can help you plan forwards and see what kind of activities you should be doing or taking up at what time of year. Perhaps your personal seasons coincide with nature's, perhaps not; there is no right or wrong, only what is right for you.

The Principle of Cause and Effect: "Every Cause has its Effect; every Effect has its Cause..."

This is the principle that there are no accidents or chance events. Every thought, word, and action creates a ripple effect that results in a consequence. It is the law of karma, where every cause has an effect, and every effect has a preceding cause. This principle reminds us that we are the masters of our own destiny through our choices.

The builders of Avebury and Stonehenge did not work by chance; they were master practitioners of this principle on a monumental scale. Their deliberate placement and alignment of stones were intentional acts of spiritual engineering, designed to create specific energetic and spiritual effects. For example, the precise solar and lunar alignments at Stonehenge were not mere observational tools—they were the *cause* designed to create the *effect* of marking and channelling powerful celestial energies during key moments like the solstices. Similarly, the vast, encompassing design of Avebury was a conscious *cause* to create the *effect* of a communal, sacred space that fostered social cohesion and a deep connection to the Earth's energies. By understanding how the landscape, celestial alignments, and geometry could be manipulated, the builders were able to apply these laws to manifest their profound visions in stone. Both sites are therefore powerful examples of applied energetic and spiritual technology, where conscious intent serves as the ultimate cause for monumental effect.

On a more personal level, this principle holds a powerful truth about our own lives: we are the master of our own destiny through our choices. Think of it this way: our life is a collection of moments and choices that have brought us to where we are right now. The choices we make now, in this present moment, become the cause that creates our future, which will, in turn, become our present tomorrow. The more aware we become, the better we know our true nature, our strengths, and our weaknesses. This increased self-knowledge allows us to make more informed choices—choices that are in greater alignment with our soul's purpose and our deepest desires.

The difference between a desired reality and an undesired one is often just a matter of a single degree. Consider the metaphor of a plane's flight path. On a short journey, if the pilot corrects the course by one degree, the impact is minimal and can be easily adjusted. However, on a long-haul flight, a one-degree deviation from London to Los Angeles could be the difference between hitting the landing strip and ending up in Mexico or the Pacific Ocean. This seemingly minor shift in direction, amplified over thousands of miles, leads to a completely different destination. This mirrors the choices we make every day. Our small, seemingly insignificant thoughts and actions accumulate over time to create a life far from our intended course—or perfectly aligned with it.

As you can see, it's not about making a single, grand decision, but about the accumulation of small, daily choices. Each time you choose to respond with patience instead of anger, to see a setback as a lesson instead of a failure, or to cultivate a positive mindset instead of a negative one, you are actively creating your future. You are shaping your life situation, big or small, through a continuous chain of cause and effect, ultimately mastering your life by mastering your choices. *This* is the subtle, ongoing work of the conscious creator.

The Principle of Gender: "Gender is in everything; everything has its Masculine and Feminine Principles..."

This principle doesn't refer to biological gender, but rather to the masculine and feminine energies that exist in all things. The masculine principle is about giving, action, and direction, while the feminine principle is about receiving, nurturing, and creating. All things and people contain a balance of both. True creation comes from the union of these two energies.

On a practical, everyday level, we can see these energies at play in our own actions. We can say that our actions have different genders. Masculine actions are outwardly directed and linear, such as going for a run, tackling a project at work, or engaging in a debate. They are about 'doing' and 'achieving'. Feminine actions are more inwardly directed and cyclical, such as meditation, journaling, or simply taking time to rest and nurture your inner self. They are about 'being' and 'feeling'. Neither is better than the other; they merely serve different purposes and are equally important for a balanced and whole life. The true art lies in knowing which energy to engage with at the right time.

The layout and individual features of both Avebury and Stonehenge might symbolise the interplay of masculine and feminine energies. The encompassing circles of Avebury could represent a receptive, nurturing energy, while the upright trilithons of Stonehenge might symbolise a more active, directional force.

The Observer Effect: Your Personal Alchemy in Action

Modern science, particularly through the framework of quantum physics, has begun to provide a language for these same Hermetic concepts. While the ancients spoke of a living, vibrating cosmos, quantum mechanics reveals a subatomic world where particles exist in multiple states at once and are influenced by the very act of observation. The ancient alchemist's belief in the power of conscious intent and inner state to influence the outer world is now finding a surprising parallel in the quantum realm.

In the quantum world, the act of observation changes reality. The double-slit experiment, a cornerstone of quantum physics, shows this beautifully. When particles are not being watched, they behave like waves, passing through two slits at once and creating an interference pattern. But the moment a detector is used to observe which slit a particle passes through, its wave-like behaviour collapses. The particle acts like a solid object, choosing one path and creating two distinct bands on the screen. This is the observer effect at its core: the act of conscious observation itself changes reality. It's a powerful metaphor for how our own consciousness can influence the external world. It also draws a parallel to demonstrate how we can "observe" our desired reality into reality if we know how to consciously recode (rewire) our neuropathways.

THE OBSERVER EFFECT

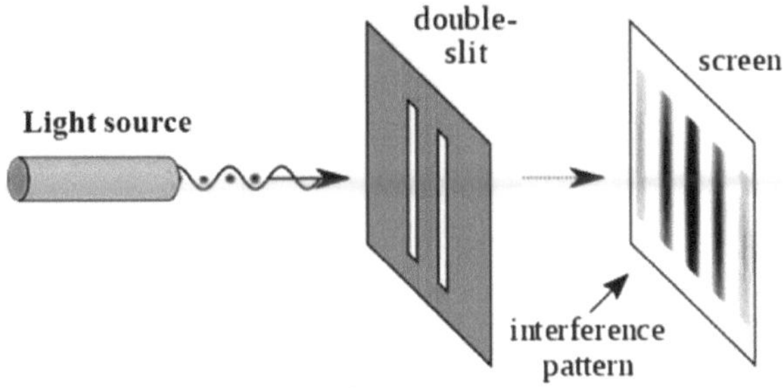

When I was a teenager, I wanted to become a published bestselling author and speaker, and this desire has always burned strongly in me. Going into that scene in my mind back then would make me feel confident, content, fulfilled, creative and motivated. I used to see myself living a bohemian lifestyle. I would be in my library in my Victorian apartment with high ceilings. It was Autumn, dark and cold outside, which didn't affect me apart from an occasional, soft draft coming through the cracked open windows, the heating was on, so I didn't feel chilly. Bookshelves filled to the rim with books would fill every available wall space from floor to ceiling. Thick, burgundy velvet curtains would cover tall windows. I was sat on a comfortable Chesterfield office chair by an old, heavy oak desk, typing away on an old-fashioned typewriter in the light of a green Tiffany desk lamp. I would have my favourite glass of deep, red wine next to me – and a cigarillo. I would write late into the night in the cosy, warm room, wearing heavy black cat eye spectacles a paisley-patterned burgundy silk morning gown, which matched the curtains. The silk felt cool and soft, caressing my forearms as my arms moved to write novels, prose and short stories. I could taste the wine on my tongue with its slight bitterness but still rich with a fruity bouquet. I could smell the musky cigarillo in the air and the slightly metallic odour of ink from the typewriter – and a hint of my Chanel No. 5. I always wore a full makeup making my chestnut-green eyes “pop”, and my long, dark hair was loose. The sounds in the room were dominated by the tapping sound when I hit the keys on the typewriter, the ping at the end of the line and the click-like swoosh when I adjusted to a new line. I was feeling good, a glowing, warm feeling inside from living my ideal existence, living a life I loved fascinatingly imbued in mystique, much like being my own muse.

Even when I think back to this today, I see this memory like a postcard on a screen behind my forehead. I have replayed it so many times it became a part of me, all the sensory impressions still oozing out of the postcard much like a 5D film (I’m in it). This is also why it was so easy for me, as a mature adult, to get back into this alignment, it was already there. Obviously, the scene has modified onwards from a teenager’s perception, but the core is still the same, namely writing books, short and long books. I am now a mature woman who don’t smoke, I rarely drink alcohol, and I am mainly writing through daytime hours. However, I DO have black, heavy cat eye reading glasses and I still adore Chanel No. 5. I AM a published author. I write books designed to help people on a more practical level rather than creating made-up stories, even though it would be so easy to let my imagination take detours. Perhaps when my other work is done, if my desire to do so is still there.

Even today when I think back to this today, this scene always puts me in a state of confidence, contentment, achievement and fulfilment, creativity and motivation. Every time I write I am in this sphere. Unknowingly to the science behind it, all those years ago I created an anchor—a memory from the future—that became the core blueprint for my life today.

My life took many and long detours from this path but when I learned how to consciously create, I realigned to my core and made it happen rather than forever seeing it as a utopian dream.

I trust you were able to step into that scene as you read it. This detailed story is the profound motive: it demonstrates how to imprint your desires so deeply that you can observe them into existence—the secret to creating future memories. Think of it like the light neutrons being observed going through the slits, and a future where you have no conscious direction is the pattern with interference. Now at the end of this chapter, I am giving you two tools that make up part of the process in becoming a conscious creator, an exercise and a guided meditation.

The Observer's Anchor: A Technique for Intentional States

Our brains are a lot like a computer: they have a unique file system for storing memories and emotions. Anchoring is a powerful technique for creating a mental and physical 'shortcut' back to a positive, resourceful state that you choose with intention.

The Power of Music and Memory

Our unconscious mind is always creating links between our emotional state and the sensory information around us—what we see, smell, hear, and feel. A great example of this is how our brains anchor emotions to music.

Think of a song that you absolutely love. Perhaps it's a tune that was playing at a special event—a first dance, a major celebration, or a fun road trip with friends. In that moment, if the emotion and event created a big impact on you, your brain links the sounds of the song to the joyful emotions you're experiencing. Every time you hear that song again, your unconscious mind recalls that original emotional reality, and you feel a rush of that same happiness or nostalgia. The song itself becomes an anchor for that positive state. This is a universal experience, and it shows how your brain is constantly, and without conscious effort, creating these shortcuts between your senses and your emotions. Other times, if there is less of an impact, your unconscious brain will need three repetitions to clearly create the link.

Just as the Observer Effect shows that our conscious observation can change reality, this exercise demonstrates how your conscious presence and physical action can instantly bring back a specific feeling.

Choose a State:

We start by identifying a positive state you want to have on tap. This could be confidence, calm, joy, or absolute focus. It's best to start with a strong emotion you can remember easily.

Choose an anchor:

An anchor is a place you choose to put a physical imprint of the emotion, or state, you are programming yourself for. It should be something you don't do often in daily life, so the anchor is strong and clear. You could gently squeeze your left thumb and forefinger together. You could tap a specific knuckle. You could briefly press your fingers against the side of your neck. You could grab your wrist and squeeze, but please don't squeeze so hard you cut off the blood flow to your hand. Do this for about 5 to 10 seconds as you hold on to the peak of the feeling.

Recall and Relive:

Close your eyes and recall a very specific memory where you felt that state intensely. It could be a moment from your childhood, a recent achievement, or a time you felt completely at peace. Re-live that memory as vividly as you can.

See what you saw. What were the colours, the shapes, the light?

Hear what you heard. What were the sounds, the voices, the music?

Feel what you felt. What was the feeling in your body? Was it a warmth, a lightness, a sense of power? Where in your body did you feel it most? Help the feeling grow and get stronger inside of you, perhaps until it fills your entire body and you feel like you become the emotion...

Set the Anchor:

When you are at the peak of that feeling—when it is most intense—make a unique, specific physical gesture or touch. Activate the anchor you chose before you started.

Break State:

Open your eyes, take a deep breath, and think about something else entirely. Shake your hands, look around the room. You want to clear your mind of the memory.

Test the Anchor:

Now, simply perform the anchor gesture you created. Consciously observe the feeling from the memory as it rushes back into your body.

How it Works:

You have created a powerful link between a physical action and an emotional state. Your brain has now stored this state under a new, easy-to-access 'file name'—the anchor itself. You can now use this anchor anytime you need to feel confident, calm, or joyful. An anchor will fade over time if you don't use it, you may have to do the process again; this is because our unconscious that created an emotional link to this physical experience likes repetition.

Practise this exercise a few times with different positive states, using different anchors for the various states, strengthening your ability to consciously call up your desired emotional reality.

Conscious Creator Meditation

At the end of this book there is an appendix with the script to the Conscious Creator Meditation that completes this chapter. To deepen your journey with "The Weave of Stone and Light," you're invited to listen to the specially crafted guided meditation that complements this script.

This meditation will help you embody the principles described in the book and connect more fully with the sacred energies around you.

Simply visit www.sirilagerlov.com (under development on publishing date) or scan the QR code and enter your email address where prompted to receive instant access to the audio file.

Now we are ready to move on to the megaliths.

CHAPTER 5

THE TWO HALVES OF THE GREAT WORK: AVEBURY AND STONEHENGE

5. The Two Halves of the Great Work: Avebury and Stonehenge

Our journey into the energetic landscape of the British Isles would not be complete without a closer look at its two most magnificent monuments: Avebury and Stonehenge. A megalith is, at its simplest, a "great stone." As we will see, while both are made of these monumental stones, they stand as two distinct expressions of ancient wisdom, each with a different energetic purpose, each embodying a different stage of our own inner alchemical process. The ancients did not just move stones; they were architects of the invisible, creating sacred spaces that mirrored the principles of the universe. The sheer, back-breaking effort they put forth to source, transport, and align these stones speaks to a profound intentionality: a commitment to a vision that we, as conscious creators, can learn from today.

Skara Brae: The Hermetic Foundation in Daily Life

We begin our journey in the far north of Scotland with Skara Brae on the Orkney Islands. While Skara Brae is celebrated as the best-preserved Neolithic village in Western Europe, offering an unparalleled glimpse into the domestic life of the Grooved Ware culture around 3200-2500 BC, its true significance lies in its role as the human heart of the Sacred Landscape of Orkney. Situated in the same ceremonial topography as the massive Ring of Brodgar and Maeshowe chambered tomb, the settlement was not just a collection of functional dwellings but an integrated part of a complex, ritual-charged environment. The uniform, meticulously designed stone houses, particularly the "altar-like" dresser opposite the entrance, suggest a deliberate connection to ancestral or cosmological order. Furthermore, the presence of House 7, with its anomalous lack of beds and evidence of possible foundation rituals (such as burials beneath the floor), challenges the purely secular view, suggesting that Skara Brae may have been home to the privileged or theocratic class who were the conscious creators and custodians of the surrounding monumental complexes. The standardised and precise layout of Skara Brae, situated in alignment with the great ritual complexes, embodies the Hermetic principle of "As Above, So Below," reflecting the belief that the patterns of the human world mirror the cosmological order and the conscious creators who established them.

The Two Hearts of Wiltshire

While this ancient village demonstrates the presence of Hermetic principles and conscious creation in the daily lives of the Neolithic people, it is in the rolling chalk downs of Wiltshire that the Great Work of the landscape truly finds its most monumental expression. Our primary focus now shifts south to the two great stone hearts of Britain: Avebury and Stonehenge.

Avebury: The Encompassing Embrace of Raw Potential

As we first turn our attention to the monumental landscapes of Avebury, let us be struck by its sheer scale. This isn't just a stone circle; it's a vast, encompassing prehistoric complex, the largest of its kind in Europe, embracing entire villages and a network of henges, avenues like the Kennet and Beckhampton, and enigmatic features like the Sanctuary and Silbury Hill. Unlike its more formally defined cousin to the south, Avebury exudes a sense of organic integration with its surroundings, a sprawling embrace of the land itself.

This monument, constructed between 2850–2200 BC, during the Neolithic, or New Stone Age. The monuments here are preserved as part of a Neolithic and Bronze Age landscape for the information they provide about prehistoric people's relationship with the land. Originally well over 100 stones, Avebury has a sad history where the villagers, encouraged by the church in an effort to discourage paganism, started removing the stones in the Middle Ages.

Avebury feels less rigidly planned and more like a continuous unfolding across the landscape. Its structure can be viewed through the Hermetic lens as a representation of the initial, broader stages of the alchemical Solve et Coagula process. It is the Prima Materia, a vast, unformed potential, symbolising the dissolving of boundaries and the encompassing of diverse elements. The sacred landscape of Avebury invites us to embrace this process of dissolving our rigid forms and opening ourselves to the vast, encompassing potential of our being.

PRIMA MATERIA

Prima Materia is an ancient alchemical term for the primitive, unformed essence of all things. It is the raw, chaotic material from which everything is created. In Aristotelian philosophy, it was the "first matter" that could be given any form. In alchemy, it was the essential starting point for the Great Work, the process of transformation and creating the Philosopher's Stone.

What does this mean for us today?

In modern psychology, particularly as explored by Carl Jung, Prima Materia is a beautiful metaphor for the chaotic and untamed potential within our own minds. It represents the "messy places" of our psyche—the unresolved traumas, the unexamined fears, the inner chaos—that signal an opportunity for transformation. Rather than being a problem to be fixed, this "messiness" is the very stuff of our genius. Just as the ancient alchemists needed to understand their raw materials to transform them, we must be brave enough to look at our inner chaos as the raw material for our greatest growth.

Stonehenge: The Focused Culmination of Intent

In stark contrast to Avebury, the iconic silhouette of Stonehenge immediately conveys a sense of focused power and deliberate engineering. The iconic stone circle was erected around 2500 BC and was altered for centuries until around 1500 BC. Its later phases, with the precisely arranged sarsen trilithons and the bluestone horseshoe, speak to a more clearly defined intention, a monumental undertaking of will and commitment.

To build this site, the ancient peoples went to extraordinary lengths. The massive sarsen stones were transported from locations like West Woods on the Marlborough Downs, a significant haul. Even more remarkably, the smaller bluestones were dragged all the way from the Preseli Hills of west Wales, a journey of about 250 kilometres (155 miles).

This superhuman effort is a tangible metaphor for the focused intent required for conscious creation. This was not a casual project; it was a profound act of will that required generations of dedication. Just as the builders laboured for years to align the stones, we must apply deliberate, sustained focus to align our internal world with our desired external reality.

Stonehenge's focused structure is perfectly aligned with the solstices, demonstrating a deliberate, clear alignment with the cosmos, an astronomical significance that suggests its use as an ancient calendar for tracking key celestial events. Hence, this speaks to the Hermetic Principle of Correspondence: "As Above, So Below." Furthermore, this powerful alignment and focused structure sit on a very prominent ley line: the St. Michael's Ley Line, which suggests the site was engineered to channel the Earth's vibrational energy with great intensity.

The strong associations of Stonehenge with solar alignments, particularly the solstices, are well-documented. This precise engineering links the 'Stone' directly to the 'Light' of the cosmos, making the site a deliberate channel for the cyclical energy of the sun. The ancient builders were weaving the terrestrial stone with the celestial Light, aligning Earth with the solar will.

STONEHENGE

Today, this focused energy continues to draw people from across the world. Both Avebury and Stonehenge have become sites of modern pilgrimage, attracting thousands of people each year seeking connection, renewal, and a link to ancient wisdom. This global magnetic pull demonstrates that the sacred power embedded in the landscape is very much alive and operating as a powerful anchor for human consciousness.

This monument's focused structure might be situated on more direct, linear power lines, channelling energy with greater intensity towards its central point. This can be viewed through the Hermetic Principle of Vibration: a focused amplification of specific energetic frequencies.

The ley lines themselves are an expression of the planet's vibrational energy. The international extensions of the ley lines even suggest a deeper, global network. Other proposed alignments include The Ridgeway Alignment, an ancient trackway. The monument's focused structure might be situated on more direct, linear power lines, channelling energy with greater intensity towards its central point. This focus on celestial precision and linear energy channels makes Stonehenge a metaphor for the focused intent required to bring your goals into reality, which is the second, crucial phase of Coagula.

Expanding the Sacred Landscape

While Avebury and Stonehenge serve as the perfect foundational metaphor for the Great Work, the energetic connection between ancient sites and human consciousness is evident across the entire British Isles. This broader network of megaliths demonstrates that the Hermetic Principles are universally anchored in the landscape, not merely in one location.

Glastonbury Tor (Somerset): This iconic, highly magnetic hill is a central node on the St. Michael's Ley Line and is deeply woven into Arthurian and Christian mythologies. It perfectly embodies the Principle of Correspondence ("As Above, So Below"), acting as a terrestrial anchor for celestial and legendary energies. Its adjacent Chalice Well embodies the receptive, 'feminine' energy of the Principle of Gender.

Castlerigg Stone Circle (Cumbria): Situated high among the rugged mountains of the Lake District, this dramatic site instantly conveys immense natural power. The placement of these stones in such a wild, powerful landscape illustrates the raw intensity of the Principle of Vibration. It invites us to consider how our own energetic frequency changes when we surround ourselves with the fundamental power of nature.

Chapter Summary: A Tale of Two Temples

Ultimately, the enigmatic stone circles of Avebury and Stonehenge represent distinct but complementary approaches to engaging with cosmic and terrestrial energies. They embody two crucial stages in our journey as a conscious creator.

Avebury signifies the vast potential and the encompassing nature of a spiritual journey. It is the courage to embrace the inner chaos, to dissolve old boundaries, and to recognise your own raw potential. This is the stage of letting go and trusting the process: Solve.

Stonehenge represents the focused culmination of that journey. It is the deliberate act of aligning with the cosmic rhythms and using your intent to manifest a specific, desired outcome: Coagula.

You cannot achieve the focused intention of Stonehenge without first embracing the raw potential of Avebury. Both sites, in their unique ways, embody the Hermetic principles of Correspondence, Vibration, and Gender, acting as tangible anchors for an ancient wisdom that is still alive today.

The enduring power of these diverse sites, stretching from the gentle hills of Somerset to the wild north of Scotland, proves that the creation of a sacred landscape was a concerted effort across ancient Britain. It provides tangible evidence that these universal laws of transformation were built into the fabric of life, waiting for the conscious creator to rediscover them today.

The Alchemist's Exercise: Solvō et Cōagulā

Before moving on, take a moment to apply the wisdom of these two temples to your current life or career goals. This is your personal exercise in Inner Alchemy:

Identify Your Avebury (The Solvō Phase): What is the chaos or Prima Materia in your current situation? What are the messy, unexamined areas of your business or personal life; the confusion, the fear, the unexpressed potential? This is the raw material you must first embrace and dissolve. Write down three elements you need to be brave enough to look at.

Identify Your Stonehenge (The Cōagulā Phase): What is the clear, focused, aligned intention you are trying to manifest? What is the one-degree course correction that will align your life with your highest purpose, requiring the monumental effort of a committed leader? Write down one singular, aligned intention.

By first identifying and embracing your chaos, you find the commitment required to then build your desired reality with focused, stone-moving intention.

CHAPTER 6

THE ALCHEMIST OF TIME: MASTERING THE NON-LINEAR

6. The Alchemist of Time: Mastering the Non-Linear

One of the greatest illusions of our modern lives is the relentless, linear march of time. We measure it in minutes and hours, in deadlines and appointments. We feel like we're either running out of it or that we're forever chasing it. But what if time, as we know it, is just one aspect of a much more fluid and malleable reality? What if we could, like a master alchemist, learn to transmute our experience of time, shifting from a linear, stressful existence to a non-linear state of being that is more aligned with our true nature?

The Zero Point (In-no-sense)

To understand this, we must first unlearn the idea that time is a fixed, unwavering line from past to future. Esoteric traditions have long taught that all moments, past, present, and future, are in some way co-existent. In this view, time is not a river, but a vast, silent ocean, where every moment is a point of interconnected potential.

This is not just a poetic idea; it resonates with the strange and beautiful truths of quantum physics, where particles exist in multiple states at once until they are observed. This observation, this act of pure awareness, your Light, is the force that collapses potential into physical reality. This is the truth that reveals itself when you are in a state of deep inner alignment, namely your 'zero' point.

Your zero point is that sacred, quiet place within you, the eye of the storm. It's a state of profound peace and innocence, or as William Whitecloud calls it, 'in-no-sense'. It is the space of pure observation, where you intentionally suspend the analytical mind's need to define, judge, or make sense of anything. Like a newborn child, you are wholly present, ready to absorb the information and potential around you without the bias of judgment or perception. In this neutrality, you are not defined by the wounds of the past or the anxieties of the future. You are simply the conscious observer of reality, and it's in this powerful, non-reactive centre that you have the power to shape it.

The Three-Breath Zero Point Anchor

Let us experience this through a practical exercise. This technique can be performed anywhere to immediately shift you out of linear, stressful time and into the potent Zero Point. You can add a physical anchoring point for this if you so wish; it is best to choose a point that is easily reached and where there are no anchors from previous. For instance, you might squeeze your left wrist with your right hand (albeit not too hard), or if you wear your watch on your left arm, you may find it better to use your right wrist.

*1. Inhale the Past (Solve):*Take a slow, deep breath and consciously acknowledge any pull from the past; any regrets, decisions, or past hurts. Exhale slowly and fully, gently releasing the emotional weight of that history, leaving the lesson but letting go of the chain. *You may wish to repeat this step up to three times until you feel a noticeable shift.*

2. Inhale the Future (Suspend): Take a slow, deep breath and acknowledge any anxiety, pressure, or expectation from the future; any deadlines or worries. Exhale slowly and fully, releasing the need to control the outcome, trusting in the unfolding. You may wish to repeat this step up to three times until you feel settled.

3. Inhale the Now (Coagula): Take a final, deeply satisfying breath and simply be. Feel the support beneath you, notice the stillness. In this breath, you realise you are perfectly neutral, perfectly safe, and perfectly present. Repeat if you need a little extra time to notice the grounded peace, however, as with everything – with practice it might suffice to do this step only once.

4. Set Your Anchor: When you feel this stillness nice and strong, the deepest sense of peace and presence, make the physical gesture you chose (e.g., squeeze your wrist). Hold the anchor for 5 to 10 seconds.

5. Break and Test: Open your eyes, take a deep breath, and actively think about something completely different—look out of your window, check to see what time it is: break the state. Now, simply reapply the anchor gesture and notice the feeling. If the feeling of peace immediately floods back when you close your eyes and think of it, your anchor is set! If not, repeat the entire three-breath process to build the association between the physical touch and the Zero Point state until it works instantly.

The initial steps often require concentration, but with time, you'll find you can "call up" the Zero Point in your mind without needing the physical trigger at all, having successfully transmuted the lead of anxiety into the gold of instant peace. In my programmes there are deep meditations to train you to get into the state of in-no-sense to enable you to consciously create anything.

Alchemists of time

The ancient builders of the megalithic sites, such as Stonehenge and Avebury, may have understood this on a very deep level. While we often see these monuments as clocks or calendars for cosmic events, or even to predict tide surges, their purpose may have been far grander. They were perhaps designed not to simply measure the passing of linear time, but to help people step out of it. By aligning themselves with the cycles of the sun, moon, and stars, and by being within the powerful energetic fields of the ley lines that flow through these sites, they could have used these locations as catalysts to enter a state of timelessness.

Think about the feeling of being at one of these sacred sites. It's a sense of profound peace and a connection that transcends the generations. In that moment, do you not feel the echoes of those who came before you? This is because in the non-linear perspective, the past is not gone; it is simply a different energetic frequency, and when you are in your zero point, you are a powerful receiver. These sites, with their sacred geometry and alignment to the cosmos, could act as spiritual antennae, helping us tune into and receive the wisdom of the past, not as a memory, but as a living energy.

This understanding of time is deeply rooted in the Principle of Rhythm, which teaches that everything flows, ebbs, and has a cycle. It's also anchored in the Principle of Correspondence: 'As Above, So Below.' The timelessness of the stars and planets is mirrored in the timelessness we can access within ourselves, if we choose to.

The Alchemist of Time is someone who understands that the real power is not in changing the past or predicting the future, but in consciously creating from the powerful, timeless now. It is the art of turning the leaden weight of our history and the golden potential of our future into the gold of our present reality. This is the ultimate act of self-mastery, a way of living that is no longer defined by the constraints of linear time, but by the boundless potential of our own consciousness. In this very moment, you have the power to observe your desired future into existence.

The true mastery of the Alchemist of Time is the ability to transmute the lead of the past into the gold of the present. You don't need to change what happened; you simply need to change the story you tell yourself about it from your zero point. From this place of neutrality, the old memory is no longer a wound but a neutral piece of information—a necessary step—which allows you to harvest the wisdom and power from that past experience, making it a source of strength, not a chain.

CHAPTER 7

CONCLUSION: THE LAUNCHPAD OF THE CONSCIOUS CREATOR

7 Conclusion: The Launchpad of the Conscious Creator

Our journey through the lens of modern alchemy culminates with the profound realisation that Avebury and Stonehenge are far more than just stone and earth. They are the Weave of Stone and Light, tangible manifestations of an ancient understanding of energy, interconnectedness, and cosmic alignment. Stone represents the matter, the body, and the Earth's grounding energy; Light represents consciousness, the cosmos, and the boundless potential of the non-linear.

The Hermetic principles we have explored provide a timeless framework for understanding reality's interconnectedness. They show us that the ancient tenets of Mentalism, Correspondence, and Vibration are not just abstract ideas from a forgotten past, but a robust system for comprehending the seamless web of existence—from mind to matter, past to present. In this grand tapestry, alchemy is revealed to be the true Art of Transformative Alignment, both inner and outer. The symbolic pursuit of turning lead into gold was never about material wealth; it was the ultimate quest for inner mastery and conscious creation, synthesising the personal with the universal.

What is perhaps most remarkable is how modern science is now echoing these ancient intuitions. Our discoveries in neuroplasticity confirm that our minds are not fixed, but fluid and transformable. The perplexing world of quantum physics similarly suggests a reality that is far more fluid, dynamic, and interconnected than we ever believed possible. As we learn more about how we and "IT" are all related, the significance of ancient sites becomes increasingly important, reminding us of timeless truths that our ancestors understood intuitively.

The Two Halves of Your Great Work

The journey through these chapters is your roadmap for inner work. Remember the two halves of the Great Work, which you now have the tools to complete:

Avebury is your Solvō: It is the courage to embrace the Prima Materia—the messy, untamed potential, the inner chaos—and dissolve the old boundaries that limit you.

Stonehenge is your Cōagulā: It is the disciplined, focused commitment required to align your will with cosmic law and bring your vision into solid form.

This brings us to your personal power. By consciously accessing your Zero Point, you step out of the illusion of linear time and the constraints of the past. From that state of 'in-no-sense,' you become the powerful observer who collapses the potential of the universe into your desired present reality. You are, quite literally, the Alchemist of Time.

A Modern Pilgrimage

These stone circles are not just archaeological curiosities; they are consciously designed nodes of energetic and cosmic resonance. They are tangible echoes of an ancient people who understood the subtle energies of the land and sky. I believe that even though they are of the past, they are just as highly important today, as facilitators of healing, guidance, and community. They help us to keep the mysteries and dreams alive.

The true magic of these sites is their capacity to serve as personal portals for transfiguration. As I have experienced in both solitude and ceremony, standing on these ancient grounds is an immediate act of healing and alignment. It is an invitation for you to join the enduring human quest to understand your place in the cosmos. Much like people of ancient and modern times visit sacred sites such as Avalon (Glastonbury, the Earth's heart chakra) or Jerusalem, these sites are sacred places of pilgrimage for significant life changes.

Ultimately, you are all conscious creators, weaving your own reality in alignment with the universe. And in this great work, you now have the guidance of ancient wisdom and the powerful resonance of the land itself. I invite you to step fully into your role as a Master Coach of your own life.

Closing Thoughts and Next Steps

The practices within The Weave of Stone and Light lead you to a state of profound inner stability. This is the foundation upon which you can build true, lasting self-belief—what I call Authentic Confidence. It is the ultimate goal of the Modern Alchemist, and it is the starting point for our next journey together...

Acknowledgements

The Weave of Stone and Light is dedicated to the ancient truth that guides us and the community that nourishes us. While I cannot possibly mention every single person who has illuminated my path, I must express my deepest gratitude. The origin of this book was sparked by a talk I did on 'Modern Alchemy and Ancient Stone Circles' for the Society of Ley Hunters. While the circumstances around that presentation were challenging - a story for another time - the immense amount of work and research I dedicated to it became the seed of this entire manuscript. Adrian Hyde, Alan Bowers, and Gaynor Joseph, I am so grateful for trusting in me and giving me that pivotal launchpad for this book project.

A big thank-you to my parents and Carolyn for reading through the manuscript and providing constructive feedback, chapter by chapter.

A big thank-you to my children Mia, Lucas, and Nic. Your patience, love, and understanding throughout the writing of this book have been truly remarkable. Bringing a work of this magnitude to completion required periods of intense focus and time away for research, networking, and deeply engaging with the material. Yet, you supported my creative journey with grace, understanding that the focus I applied was necessary to weave together the Stone and Light of this message.

I am especially grateful for your curiosity—it was Nic's simple question, "Where are the stones from?" upon discussing the enigmatic stone circles, that made me realise I had omitted a crucial piece of information from the script. This book is for all of you. You are all so gorgeous, clever, I'm so proud of you, and I love you all to the moon and back.

To mentors and masters of modern alchemy and coaching, I send infinite gratitude for showing me how to align with my innermost values and align my microcosmos to my macrocosmos, teaching me so much about practical alchemy—and ultimately life. Christopher M. Duncan, a big thank you for generously letting me use the models for the Double Slit Experiment, the Levels of Consciousness and the ones I have combined into the Applied Hermeticism model.

And gratitude to Alvin for introducing me to Hermeticism back then.

To all of you who have added to this life's tuition, be it as a teacher, a coach, mentor, friend or partner, I thank you from the bottom of my heart.

My deepest thanks go to Graham Phillips; Author and Expert on Earth Mysteries, Philip Carr-Gomm; Former Chief of the Order of Bards, Ovates & Druids (OBOD) and author of 'Druid Mysteries', Peter Knight; Author, Researcher, Dowsing and Earth Mystery Expert and Co-Founder of Dorset Earth Mysteries Group, Adrian Hyde; secretary of the Society of Ley Hunters and Hannah Knies; Creative Development Coach & Co-Founder of Magnetic Mind Certification, for their generous endorsements and insightful contributions to the back cover and the start of this book. Their belief in this project means the world to me.

Appendix: The Conscious Creator Meditation Script

The Conscious Creator Meditation

This guided journey is your practical step into the Weave of Stone and Light. Throughout the book, we explored the concept of the Zero Point, or in-no-sense if you like, that sacred centre of balance where all potential resides. This meditation is designed to show you what you can create there.

Think of this not as a passive relaxation exercise, but as an active alignment. We will use gentle guidance to help you step out of linear time and into the timeless reality of the Alchemist of Time. Allow the spoken words to anchor you in the present, so you can consciously weave the Light of your intention into the Stone of your physical reality.

Your great work observing your future reality into manifestation with the help of your Superconscious begins now.

Take a comfortable seat or lie down.

Allow your body to settle,

and feel a sense of support beneath you, the seat, the bed, just notice that you are safe and supported.

Gently close your eyes.

Begin by taking three slow, deep breaths.

With each inhale, breathe in a sense of calm and peace.

With each exhale, release any tension or stress.

Inhale peace...

Exhale tension.

Breathe in peace

Breathe out worries

Breathe in peace

Breathe out discomfort

Inhale peace

Exhale tension

Know that there is nowhere you need to go, there is nothing you need to do, you are just here, observing, breathing, feeling and living.

Imagine that you are walking along a path going through a large, open glade in the forest.

The path continues into a forest, and you follow it.

You feel safe and curious as to where the path takes you, and to your surroundings.

Notice all the details around you as you walk.

What can you hear?

Are there birds here, chirping and singing?

What kind of trees are those on either side of the path?

Can you smell the forest ground? Is it full of ferns, old leaves?

Can you see any animals?

You just notice as you walk along the path.

After a little while, there is a fork in the path, and you choose the left.

You continue walking along at your own pace and continue with your curious perception of everything around you.

After a little while, the forest opens, and you step into an open landscape.

You get blown away; the path continues almost straight ahead through a field of vibrant, beautiful purple flowers.

You almost do a little dance or pirouette or take a couple of skipping steps to take the vast ocean of purple flowers in.

As far as you can see, there are just myriads of myriads of purple flowers.

Wave after wave in the landscape.

They are a delight to your heart, they are a delight to your eyes, they are a delight to smell,

and they are a delight to touch, soft and aromatic.

You continue forward on your path and become aware of a building up ahead.

It looks like an old temple, from long before any religion.

You move closer and you notice that it looks sacred but inviting.

You walk closer and closer.

When you are close enough, you see that there are seven steps up to the entrance platform.

Each of the steps has a colour.

You walk up the steps one at the time and take a slow breath in and fill yourself with that colour before moving on to the next step.

The first step is red, you breathe in the colour and feel how it intensifies at the bottom of your spine until your root chakra glows like a red star and is activated.

The second step is orange, you breathe in the orange vibration and notice your lower abdomen shine like a bright, orange orb in your sacral chakra. This chakra is now activated.

The third step is yellow, you breathe in the yellow energy to the bottom of your sternum and feel how your solar plexus is glowing yellow and the chakra is activated.

You move on to the fourth step which is green, and as you inhale you notice how a warm green glow is intensifying in your chest. Your heart chakra is now activated.

The fifth step is light blue, you breathe in and notice the light blue spreading from your Adam's apple and a bright ball of light shines and sparkles and your throat chakra is activated.

The sixth step is indigo and you notice the vibration of this intense colour in your forehead, just above your brow. It shines brighter and brighter and your third eye chakra is now activated.

As you stand on the seventh, violet step, you feel a gentle warmth just above your head. You see a violet light orb emerge and it glows brighter and brighter.

This is your crown chakra and it is now activated.

Imagine a soft, warm light beginning to glow there, mixing with the violet. It is a golden, white light that almost sparkles, illuminating the space around you.

This is your superconscious light, your source of unlimited potential.

As you notice it, it becomes brighter and the orb shines like a little star.

As you are embraced by the warmth of this energy you notice that your mind feels clear.

You are present in the now, you are neutral.

There is only the present moment, and you notice there isa curiosity inside of you.

A pure curiosity as if you're seeing, hearing and experiencing everything for the first time.

Now, I want you to imagine a door in front of you.

This is the door to your future, your desired reality.

There is no rush; just notice the door.

What is it made of?

How does it feel to look at it?

What colour is it?

When you are ready, gently turn the handle and step through.

As you do, you step into a beautiful, vibrant reality.

The sphere of light at your crown begins to move, it floats down to illuminate your third eye behind your brow, moves on to your throat and settles in your chest where your heart centre is.

You notice as it warms your chest and grows bigger and bigger until you are wrapped in a wonderful feeling of love and bliss.

Take a moment to look around.

You have stepped into a place in your consciousness that is your desired reality.

This is your future, and you are about to find the life you love, created by you – as a conscious creator.

Here you will see what is in the life you love, the life you have created, bit by bit, choice by choice.

What do you see?

When a new thing pops up in your mind, you see it manifest in this world.

Notice the details of this world.

Are you indoors or outdoors?

What does the space look like?

Notice the colours, the shapes, the textures...

Who is there, who is in your life?

What else is there to see?

Now, notice what you hear.

Are there sounds of nature?

The gentle hum of conversation?

The sound of success or joy?

What do you feel in this future?

What is the feeling in your body?

Perhaps a feeling of profound peace, of deep authentic confidence, of immense gratitude?

Joy? Love?

Feel what it feels like to have already achieved your goal,

Feel what it is like to be the person you have become,

How do you feel inside, when you use all your senses to take in the life you have created for yourself – the life you love..

How does it feel to live the life you love?

Notice the people you interact with; notice the activities you do.

Notice what your life is like.

The perfect average day in the life you love.

What are you wearing?

What is your posture like?

What are you doing in this moment?

What does this really feel like?

Take a moment to absorb this reality.

Breathe it in.

This reality is not in some distant future; it is already a potential in your superconscious field.

It is waiting for you to observe it into existence.

Take a moment to take it all in.

Now, begin to bring that feeling, that sense of love, peace, confidence, and gratitude, back with you.

Carry it in your heart as you step back through the door.

Gently, you notice your chakras, or energy centres, going back to their normal vibration.

Starting from the crown, just above your head, you notice the violet energy ball dim ever so slightly until the chakra is back to its normal intensity.

The feeling of the life you love is still with you.

The indigo light just above your brow dims ever so gently until your third eye chakra is back to normal vibration.

The feeling of the life you love is still encompassing your very being.

Your throat is still intensely light blue, and you notice how this chakra also dims to normal activity.

The feeling of the life you love is still as strong as before.

The light from the green sphere in your heart chakra is softening, until it reaches normal working intensity.

The feeling of the life you love is still intense.

The yellow sphere in your solar plexus chakra seems to slow and dim, and the yellow light settles at normal vibration.

You can still feel the vibration of the life you love.

The orange orb in your lower abdomen also dims, and your sacral chakra is soon at normal intensity.

The feeling of living the life you love is still soaring inside of you.

The red star at the bottom of your spine softens until your root chakra is back to normal.

The feeling of the life you love is still prevalent.

Take one more deep breath, bringing the feeling of the life you love living all the way back into your body, into the present moment.

And when you are ready, gently open your eyes.

You have now planted a new seed in your unconscious.

This is the beginning of your conscious creation.

You are a master of your own reality.

Now that you know what it feels like, looks like, smells like, what it is like,

you can easily visit this field, this morphogenic field that you have created, every morning to stay aligned so you can observe it into existence.

If you come back to this meditation later,

You may see how the sacred space, or the morphogenic field, that you created, evolves and how the details are developing.

As you become more familiar with it and you take aligned action after aligned action towards the life you love, the stronger the feeling gets, the stronger the field becomes!

Because every aligned choice you are making is taking you a step closer to the Life You Love!

I wish you all the best on your creator's journey.

Happy creation day!

Much love, Coach Siri – the Modern Alchemist

Glossary

Alchemy (Modern): The philosophical and practical process of inner transformation. It is the art of consciously transforming the "lead" (limiting beliefs, unconscious patterns) of the inner world into "gold" (purpose, alignment, and conscious creation) in the outer world.

Anchor: A technique, often from Neuro-Linguistic Programming (NLP), that creates a reliable mental and physical shortcut to a specific emotional or resourceful state. An anchor links a unique, intentional gesture (like squeezing the wrist) to a chosen feeling (like confidence or peace).

Authentic Confidence: An unshakeable, purposeful sense of self that is the natural result of alignment achieved at the Zero Point. It is the empowered state from which the Modern Alchemist intentionally creates their outer reality.

Conscious Creator: An individual who moves beyond simply reacting to circumstances and intentionally designs their internal and external reality by aligning their thoughts, emotions, and actions with a desired outcome. The modern alchemist.

Consciousness (Three Levels): The map of the inner world: Self-Conscious (the Captain/logical processor); Unconscious (the Crew/keeper of emotions and programming); and Superconscious (the Guidance System/wellspring of intuition and creativity).

Dowsing: The ancient practice of using simple tools (L-shaped rods, pendulums, or forked sticks) to detect and locate subtle energies, including water, minerals, and the currents of ley lines. It is an intuitive, energetic dance amplified by the human body.

The Great Work: The ultimate goal of alchemy, referring to the entire process of spiritual and personal transformation—the journey from the base state of Prima Materia to the final state of The Philosopher's Stone (inner mastery).

Hermeticism: A philosophical tradition based on the writings attributed to Hermes Trismegistus. It is the source of the Seven Hermetic Principles (e.g., Mentalism, Correspondence, Vibration) that describe the universal laws of existence.

Ley Lines: Invisible, energetic channels or arteries and veins of the Earth that flow across the landscape, often connecting ancient sacred sites like stone circles and temples. They are believed to carry telluric energy and act as a vast planetary energetic web.

Meditation: A practice involving techniques used to train attention and awareness, and to achieve a mentally clear and emotionally calm state. It is a key tool for accessing the Zero Point.

Megalith: A word that comes from Ancient Greece. It literally means Great Stone (mega=grand and lithos=stone). In this setting a megalith is a structure, usually ancient, made of one or several large stones. In this book, sites like Avebury and Stonehenge are used as tangible metaphors for understanding ancient wisdom and energy.

Neuro-Linguistic Programming (NLP): A communication and change methodology focused on how the mind (neuro), language (linguistic), and behavioural patterns (programming) interact. Techniques like Anchoring are used to quickly change internal states.

Observer Effect: The phenomenon in quantum physics where the act of conscious observation influences or changes the outcome of a physical process. It serves as a powerful metaphor for how our consciousness actively shapes our reality.

Prima Materia: An alchemical term for the "First Matter." It symbolises the raw, untamed potential and unexamined chaos within our current state that must first be embraced and dissolved (solvō).

Shadow: A term popularised by Carl Jung referring to the parts of our personality we unconsciously repress or reject. It holds not only our inner critics and insecurities but also our dormant power, untapped talents, and brilliance. Shadow work is the process of integrating this rejected self.

Solvō et Cōagulā: Latin for "Dissolve and Coagulate", the core operational principle of alchemy. Solvō (dissolving) refers to breaking down old, fixed beliefs (the "ice") and releasing them into fluid potential. Cōagulā (coagulating) refers to reforming that potential into a new, aligned, and solid reality (the "gold").

Telluric Energy: Energy originating from the Earth (Tellus). It includes natural magnetic and electrical currents and the subtle, life-giving energies believed to flow along ley lines.

The Weave of Stone and Light: The central metaphor of this book. Stone represents the physical world, the body, Earth's grounding energy, and the physical sacred sites. Light represents consciousness, the cosmos, boundless non-linear potential, and the focused intent of the conscious creator.

Zero Point (In-no-sense): A state of deep inner alignment, neutrality, and profound peace. It is the eye of the storm where the analytical mind is suspended ("in-no-sense"), allowing the individual to become the pure, non-reactive observer who can consciously influence and shape reality.

References

This book is philosophical and metaphorical, but the core concepts are built upon foundational works and academic research.

The following foundational works, academic concepts, and online sources informed the core concepts, research, and terminology presented in this book.

Books

Bartlett, R. (2007). Matrix Energetics: The Science and Art of Transformation. Energy Psychology Press.

Bartlett, R. (2009). The Physics of Miracles: Tapping Into the Field of Consciousness Potential. Hay House.

Broadhurst, P., & Miller, H. (1995). The Sun and the Serpent: A Journey of Discovery Through the British Landscape, Its Mythology, Ancient Sites and Mysteries. Pendragon Press.

Coelho, P. (1993) The Alchemist, Harper Collins.

Doty, J.R. (2016): Into the Magic Shop. Avery.

Dyer, W. W. (2004). The Power of Intention: Learning to Co-create Your World Your Way. Hay House.

Duncan, C. M. (2021) You're Not Broken, (2023) The Superconscious Path, (2024), Integrate Your Shadow. (2025) Superconscious Transformation. Duncan Publishing.

Hicks, E., & Hicks, J. (2006) The Law of Attraction: How to Make It Work For You - The Teachings of Abraham. Hay House

Initiates, Three (Centenary Edition 2018)). The Kybalion: Hermetic Philosophy. (Note: Often published pseudonymously or anonymously, W. W. Atkinson is the widely attributed author.)

Redfield, J. (1993) (The Celestine Prophecy and further books). Satori Publishing, Hoover, Alabama.

Radin, D. (2018). Real Magic: Ancient Wisdom, Modern Science, and a Guide to the Secret Power of the Universe. Harmony.

Streicher, C. (2017) Abundance on Demand: Five Easy Steps to Master the Inner Game of Money. Results Press.

Tiller, W. A., Tibble, W. E., Jr., & Fandel, J. G. (2005). Some Science Adventures with Real Magic. Pavior.

Whitecloud, W. (2009) The Magician's Way. New World Library. (2019) Natural Success. Animal Dreaming Publishing. (2012) The Last Shaman. Wizdom Press.

Online Sources and Archaeological References

Britannica (Encyclopaedia Britannica). "Britain: Ancient and Early Medieval."

https://www.britannica.com/search?query=Britain%3A+Ancient+and+Early+Medieval.

Britannica (Encyclopaedia Britannica). “Sacred Places”

https://www.britannica.com/topic/sacred-place

Earth Sanctuary. "Ley Line Sculpture" and "Sacred Space Sculpture Garden." Earth Sanctuary. Earth Wisdom Art Science.

https://earthsanctuary.org/#sculpturegarden

English Heritage. “An Introduction to Prehistoric England (Before AD 43)”

https://www.english-heritage.org.uk/learn/story-of-england/prehistory/

Historic Environment Scotland “Skara Brae History”. *https://www.historicenvironment.scot/visit-a-place/places/skara-brae/history/*

IFL Science. "Physicists Find A Way Around Heisenberg's Uncertainty Principle, One Of The Most Frustrating Concepts In Physics.”

https://www.iflscience.com/physicists-find-a-way-around-heisenbergs-uncertainty-principle-one-of-the-most-frustrating-concepts-in-physics-81004

NASA Science. "Planetary Alignments and Planet Parades." NASA Science Solar System Exploration.

https://science.nasa.gov/solar-system/skywatching/planetary-alignments-and-planet-parades/#:~:text=Late%20April%20to%20Late%20August,visible%20at%20once%20before%20sunrise

UNESCO World Heritage Centre. "Stonehenge, Avebury and Associated Sites." UNESCO World Heritage Convention.

https://whc.unesco.org/en/list/373/

Visit Herefordshire. "The Golden Valley Pilgrim Way."

https://www.visitherefordshire.co.uk/see-do/routes-road-trips/golden-valley-pilgrim-way

www.ingramcontent.com/pod-product-compliance
Lightning Source LLC
LaVergne TN
LVHW051129210926
848482LV00001B/28

9781919215525